BATS

Emma Bassier

DiscoverRoo
An Imprint of Pop!
popbooksonline.com

abdobooks.com

Published by Pop!, a division of ABDO, PO Box 398166, Minneapolis, Minnesota 55439. Copyright © 2020 by POP, LLC. International copyrights reserved in all countries. No part of this book may be reproduced in any form without written permission from the publisher. Pop!™ is a trademark and logo of POP, LLC.

Printed in the United States of America, North Mankato, Minnesota.

102019
012020

THIS BOOK CONTAINS RECYCLED MATERIALS

Cover Photo: David Havel/Alamy
Interior Photos: David Havel/Alamy, 1; Merlin D. Tuttle/Science Source, 5, 16 (bottom); National Park Service, 6; iStockphoto, 7 (flowers), 7 (bat), 8–9, 13, 14, 17 (top), 17 (bottom), 22, 23; Shutterstock Images, 11, 15, 16 (top), 19, 20–21, 25, 26, 28, 29; MerlinTuttle.org/Science Source, 12; USFWS/Science Source, 27

Editor: Connor Stratton
Series Designer: Jake Slavik

Library of Congress Control Number: 2019942487

Publisher's Cataloging-in-Publication Data

Names: Bassier, Emma, author.

Title: Bats / by Emma Bassier

Description: Minneapolis, Minnesota : Pop!, 2020 | Series: Pollinators | Includes online resources and index.

Identifiers: ISBN 9781532165924 (lib. bdg.) | ISBN 9781532167249 (ebook)

Subjects: LCSH: Pollinators--Juvenile literature. | Bats--Behavior--Juvenile literature. | Nocturnal animals--Juvenile literature. | Sensory biology--Juvenile literature. | Animals--Flight--Juvenile literature.

Classification: DDC 599.4--dc23

WELCOME TO DiscoverRoo!

Pop open this book and you'll find QR codes loaded with information, so you can learn even more!

Scan this code* and others like it while you read, or visit the website below to make this book pop!

popbooksonline.com/bats

TABLE OF CONTENTS

CHAPTER 1
FURRY FLYER

The moon shines in the night sky.

Suddenly, a dark spot flashes in front

of the moon. A bat is flying across the

Arizona desert. It lands on a saguaro

cactus. The bat sticks its nose in the

WATCH A
VIDEO HERE!

A lesser long-nosed bat drinks nectar from a saguaro cactus flower.

cactus's flowers. It sips **nectar** from

deep inside.

Yellow pollen covers the body of a lesser long-nosed bat.

Many bats visit flowers to find food.
The flowers have **pollen**. This pollen
sticks to the bats' furry bodies. The bats
carry it to new places as they fly. They
spread the pollen from flower to flower.

BAT POLLINATION

A bat drinks nectar from a flower. Pollen gets on the bat's body. The bat flies to another flower. Some of the first flower's pollen falls into the second flower.

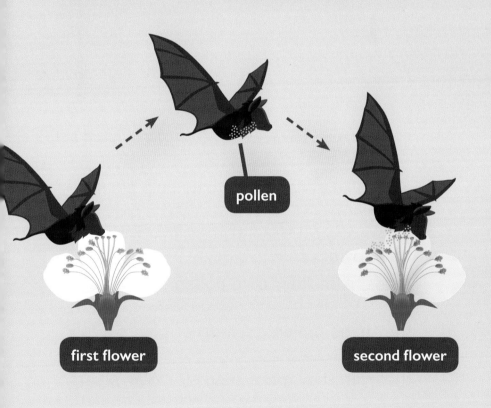

pollen

first flower

second flower

Pollinators such as bats help plants create seeds. Many plants cannot make seeds from their own pollen. Instead, they need pollen from other plants. Only then can they make seeds for new plants.

Dandelion seeds float well in the air. New dandelion plants grow where they land.

Insects spread pollen too. But some

plants rely on bats to spread their pollen.

DID YOU KNOW? More than 300 types of fruit depend on bats for pollination. These fruits include bananas, figs, and mangoes.

CHAPTER 2
BIG BATS, SMALL BATS

Bats are flying **mammals**. More than 1,300 types of bats live on Earth. All these types can be split into two main groups based on size. These groups are microbats and megabats.

LEARN MORE HERE!

Bats often have black or brown fur. But some types of bats have orange or white fur.

A microbat can have a wingspan

of just a few inches. A megabat's wings

can stretch up to 6 feet (1.8 m) long.

*The smallest bats weigh less
than a penny.*

A bat's wings are made of soft, thin skin.

The skin is stretchy and strong. The

wings connect to the bat's back and legs.

Approximately 30 percent of bats eat fruit and nectar.

Megabats eat fruit and **nectar**. They use their strong eyesight to find food.

DID YOU KNOW?

Some bats have large ears. Their ears can be more than five times bigger than their head.

WHAT IS ECHOLOCATION?

Echolocation uses sound waves to locate objects. First, a bat squeaks. The squeak's sound waves travel through the air. These waves hit objects around them, including plants and insects. Then the waves bounce back to the bat. The bat uses these echoes to find food or tell which direction to fly.

Microbats eat mostly insects. Unlike megabats, they hunt using a sense called echolocation.

LIFE CYCLE OF A BAT

A female bat often has one baby per year. The baby is called a pup.

Bats give birth upside down. They catch the pup with their wings as it falls.

A mother bat teaches her pup to fly and find food. A pup can fly after three to six weeks.

During the winter, some bats migrate. They fly to warmer areas. Other bats hibernate. This means they rest through the cold months.

Bats live for an average of five years. Some live up to 30 years.

Bats live all around the world. They can be found in many countries and many **habitats**. These habitats include rain forests, mountains, and coasts.

COMPLETE AN ACTIVITY HERE!

At least 219 types of bats live in Indonesia. That is more than any other country.

Some bats live alone. But many live in **colonies**. The place a bat lives is called its roost. A roost could be a cave, a tree, or even an attic. Bats sleep in

In the Philippines, one cave's walls are covered by an average of 60 bats every 1 square foot (0.09 sq m).

their roost during the day. They hang

upside down to rest.

DID YOU KNOW?

Bracken Cave in Texas is home to the largest bat colony on Earth. More than 15 million bats live inside.

At night, bats leave their roost

to search for food. Certain plants attract

bats more than others. For example,

A flower often smells only after it opens up.
This smell may tell pollinators that the flower's
nectar is ready to eat.

Bats tend to visit flowers with pale colors. Moonlight helps pale colors stand out in the dark.

some flowers smell like garlic, rotten

leaves, or sour milk. These strong scents

smell good to bats.

CHAPTER 4
SAVING BATS

Bats are important in nearly all environments on Earth. For example, many **tropical** plants depend on bats. These plants often grow on islands separated by the ocean. Bats can carry

LEARN MORE HERE!

Durians are a popular fruit in Southeast Asia. Without bat pollinators, durians might not survive.

pollen long distances. Insect pollinators

are too small for this job.

The Mauritian flying fox is endangered. This type of bat lives on a tropical island east of Africa.

However, many types of bats are

endangered. They may die off forever.

Some bats are at risk of losing their

habitats. But white-nose syndrome

is the biggest threat to bats. This disease damages their wings. It also causes bats to lose body fat. Bats need fat to survive, especially during winter.

The little brown bat is one type of bat affected by white-nose syndrome.

When bats eat pests, farmers do not need to use as much pesticide to protect their crops.

Without bats, more pests would

harm plants and humans. Many pests ruin

crops. Other pests, such as mosquitoes,

DID YOU KNOW?

One type of bat can eat 1,200 insects in one hour.

The lesser long-nosed bat used to be endangered. By 2018, people had helped its numbers become stable.

spread illnesses. Bats help control these

pests by eating them. By protecting bats,

humans can help all kinds of life on Earth.

MAKING CONNECTIONS

TEXT-TO-SELF

Have you seen bats before? If so, what kinds?
If not, where might you find them?

TEXT-TO-TEXT

Have you read books about other mammals?
What do they have in common with bats?
How are they different?

TEXT-TO-WORLD

Many bats are endangered. What might be
different if some bats were not around?

GLOSSARY

colony – a group of similar animals that live together in one place.

habitat – the area where an animal normally lives.

mammal – a type of animal that has hair or fur and feeds milk to its young.

nectar – a sweet, sugary liquid made by a plant.

pollen – fine, dust-like stuff that flowers create and use to reproduce.

tropical – having to do with a place where the weather is usually warm and wet.

INDEX

ONLINE RESOURCES

popbooksonline.com

Scan this code* and others like it while you read, or visit the website below to make this book pop!

popbooksonline.com/bats

*Scanning QR codes requires a web-enabled smart device with a QR code reader app and a camera.